ICML 55.0 – Optimized Lubrication of Mechanical Physical Assets Overview

ICML 55.0 – Optimized Lubrication of Mechanical Physical Assets Overview

The International Council for
Machinery Lubrication (ICML), USA

Senior Editor
Kenneth E. Bannister

NEW YORK AND LONDON

Published 2023 by River Publishers
River Publishers
Alsbjergvej 10, 9260 Gistrup, Denmark
www.riverpublishers.com

Distributed exclusively by Routledge
605 Third Avenue, New York, NY 10017, USA
4 Park Square, Milton Park, Abingdon, Oxon OX14 4RN

*ICML 55.0 – Optimized Lubrication of Mechanical Physical Assets
Overview / International Council on Machinery Lubrication (ICML) –
Senior Editor: Kenneth E. Bannister.*

Routledge is an imprint of the Taylor & Francis Group, an informa business

ISBN 978-87-7004-033-4 (paperback)
ISBN 978-10-0382-447-3 (online)
ISBN 978-1-032-65823-0 (ebook master)

Contents

Foreword   vii

Acknowledgments   ix

List of Figures   xi

List of Tables   xiii

List of Abbreviations   xv

1 Introduction   1
   1.1   Purpose   1
   1.2   Relationship with Other Standards   2
   1.3   Target Audience for this Standard   2
   1.4   Benefits of this Standard.   2
   1.5   Disclaimer   3
   1.6   Lubrication Program Management – Overview and
         Terminology   3
         1.6.1   Scope   3

2 Lubrication Program Management   5
   2.1   General   5
   2.2   Benefits of a Lubrication Management Program   6
   2.3   Assets   6
   2.4   Lubrication Program Management   7
         2.4.1   General   7
         2.4.2   Program fundamentals.   7
   2.5   Relating the Lubrication Management Program to Asset
         Management.   8
   2.6   The Lubrication Management System (LMS).   10
   2.7   Organization Plans and Objectives   10
   2.8   Lubrication Management Plan (LMP)   10

2.9   Lubrication Plans   12
2.10 Lubrication Policy   12
2.11 Development Planning and Execution   12
2.12 Implementation of Lubrication Plans   12
2.13 Performance Evaluation and Improvements   12
2.14 Typical ICML 55.1 Audit Preparation Process   13

3 Terms and Definitions   **15**
3.1   Asset   15
3.2   Asset Life Cycle   15
3.3   Asset Portfolio   15
3.4   Asset System   15
3.5   Audit   15
3.6   Capability   16
3.7   Conformity   16
3.8   Continual Improvement   16
3.9   Corrective Action   16
3.10 Documented Information   16
3.11 External Audit Assessor   16
3.12 Incident   16
3.13 Internal Audit Assessor   16
3.14 Lubrication Management   16
3.15 Lubrication Management Plan (LMP)   17
3.16 Lubrication Management System (LMS)   17
3.17 Monitoring   17
3.18 Nonconformity   17
3.19 Objective   17
3.20 Organization   17
3.21 Organizational Objective   17
3.22 Organizational Plan   17
3.23 Performance   18
3.24 Policy   18
3.25 Process   18
3.26 Requirement   18
3.27 Risk   18
3.28 Stakeholder   18
Bibliography   18

Index   **19**

Foreword

The International Council for Machinery Lubrication (ICML) is a vendor-neutral, not-for-profit organization founded in 2001 to serve the global industry as the world-class authority on machinery lubrication that advances the optimization of asset reliability, utilization, and costs.

ICML is an independent organization consisting of both paid professional staff and volunteer committees. It is a CERTIFICATION body serving both corporations utilizing lubrication assets and industrial lubrication and oil analysis practitioners worldwide; a technical AWARDS body recognizing companies and individuals that excel in the field of lubrication; a MEMBERSHIP body; and is the developer of ICML 55® STANDARDS for lubricated asset management.

Established in part to address a clear need for specific standards in all areas of lubrication management that include lubricant selection, application, training and certification, ICML has always supported such activities at ASTM, ISO and other organizations. Following publication of the ISO 55000 "Asset Management" standard in 2014, ICML marshalled its own worldwide team of 45 technical experts to develop a highly tactical, lubrication-specific standard to supplement the more general ISO document. The result is collectively known as ICML 55, an international lubrication standard that spells out the requirements and guidelines to establish, implement, maintain, and improve consistent lubrication management systems and activities.

Any trade name used in ICML 55.0:2023 is for information purposes only and does not constitute an endorsement by ICML.

The ICML thanks you for purchasing this copy of ICML 55.0:2023 Standard.

Foreword

Acknowledgments

Whenever a book, article, or paper is published, it is done so through the efforts of many different people assisting and guiding the author(s) through the entire conception-to-publication journey. ICML 55.0 is no different, and ICML thanks the following for their patience, dedication, and hard work in making ICML 55.0 a reality.

The ICML thanks all individuals who contributed suggestions, ideas, and criticisms concerning this ICML 55.0 element of the ICML 55® Standard series. The ICML thanks all individuals and corporations who have graciously granted permission for use of their copyrighted materials that include charts, diagrams, and photographs used to accompany the ICML 55.0 text.

Furthermore, the ICML also extends thanks to its dedicated staff members who have assisted in developing the ICML 55 series and to the hard-working staff at River Publishers who helped make this book a reality. ICML particularly acknowledges and extends special thanks to the following individuals for their contributions to this ICML 55.0 text:

Senior Editor, Author: Bannister, Kenneth E.

Peer Reviewer: Fitch, James C.

List of Figures

Figure 1   Corporate/system relationship.   9
Figure 2   Lubrication-related bearing/machine failure.   9
Figure 3   Key elements of a lubrication management system. .   11

List of Tables

Table 1   Typical ten-step ICML 55.1 certification process. . .   13

List of Abbreviations

ACA	Apparent cause analysis
ASTM	American Society for Testing and Materials
AW	Anti-wear
BAT	Best available technologies
BOK	Body of knowledge
BOM	Bill of materials
BS&W	Bottom sediment and water
CARRS	Classification and records retention system
CBM	Condition-based maintenance
CC	Carbon credits
CM	Condition monitoring
CMMS	Computerized maintenance management system
DF	Detectability factor
DMS	Document management system
DOE	Department of Energy
DOK	Domain of knowledge
EAM	Enterprise asset management
ECHA	European Chemicals Agency
EFL	Environment friendly lubricant
EHD	Elastohydrodynamic
EP	Extreme pressure
EPA	Environmental Protection Agency
FIFO	First in, first out
FMEA	Failure mode and effects analysis
FMECA	Failure mode effects and criticality analysis
FR	Fire-resistant
FRACAS	Failure reporting, analysis, and corrective action system
FRN	Fault risk number
FRP	Facility response plan
FTA	Fault tree analysis
FTIR	Fourier transform infrared
GHS	Global harmonized system

GNP	Gross national product
HES	Health, environment and safety
HFRR	High-frequency reciprocating rig
IBC	Intermediate bulk container
ICML	International Council for Machinery Lubrication
ICP	Inductively coupled plasma
IIoT	Industrial Internet of Things
IoT	Internet of Things
ISO	International Organization for Standardization
JIT	Just-in-time
KPI	Key performance indicator
LIMS	Laboratory information management system
LLA	Laboratory lubricant analyst
LMP	Lubrication management plan
LMS	Lubrication management system
LOER	Lubrication operation effectiveness review
LOF	List of failures
LOFM	List of failure modes
LOTO	Lock out-tag out
LSV	Linear sweep voltammetry
MIT	Massachusetts Institute of Technology
MLA	Machine lubricant analyst
MLE	Machinery lubrication engineer
MLT	Machinery lubrication technician
MOU	Memorandum of understanding
MRO	Maintenance, repair, and overhaul
MTBF	Mean time between failures
MTTF	Mean time to failure
NLGI	National Lubricating Grease Institute
OCME	Overall condition monitoring effectiveness
ODI	Operator-driven inspection
OEM	Original equipment manufacturer
OMC	Overall machine criticality
OSHA	Occupational Safety and Health Administration
PAG	Polyalkylene glycol
PdM	Predictive maintenance
PET	Polyethylene terephthalate
PF	Potential failure, also known as P-F
PM	Preventive maintenance
PPE	Personal protective equipment

QR	Quick response (code)
R&O	Rust and oxidation
R&R	Repeatability & reproducibility
RACI	Responsible, accountable, consulted and informed
RCA	Root cause analysis
RCFA	Root cause failure analysis
RCM	Reliability-centered maintenance
RDE	Rotating disc electrode
REACH	Registration, evaluation, authorisation and restriction of chemicals
RFID	Radio frequency identification
ROI	Return on investment
RPN	Risk priority number
RPVOT	Rotary pressure vessel oxidation test
RTF	Run-to-failure
RUL	Remaining useful life
SDS	Safety data sheet
SEM	Scanning electron microscope
SLA	Service level agreement
SOP	Standard operating procedures
SPCC	Spill prevention, control, and countermeasure
SSS	Spares, storage, and standby
SVHC	Substances of very high concern
SWOT	Strengths, weaknesses, opportunities and threats
TBN	Total base number
TDS	Total dissolved solids
TPM	Total productive maintenance
TSCA	Toxic Substances Control Act
TSEA	Task safety and environmental analysis
UIN	Unique identification number
VGP	Vessel general permit

1

Introduction

When it comes to mechanical equipment, best practice maintenance organizations have long recognized the positive effect that an engineered and well-balanced lubrication management program can have on asset reliability, availability, maintainability, throughput (work), quality, safety, energy reduction, carbon footprint reduction (sustainability), and cost/profits.

1.1 Purpose

The purpose of this ICML 55.0:2023 international standard is to provide an overview of a lubrication management system and processes applicable to the effective management of physical assets related to lubrication, its principles, and terminology. It also provides the context for ICML 55.1 (audit requirements) and the ICML 55.2 Guideline for the Optimized Lubrication of Mechanical Physical Assets.

It is intended, but not warranted, that this document is structurally harmonized with the international standard ISO 55001, as amended, and its subparts.

This standard identifies and defines the need for the use of well-established best practices that are applicable to a wide range of lubricated mechanical assets.

1.2 Relationship with Other Standards

This standard is intended as a companion document to be used in association with the ICML 55.1 and ICML 55.2 standards, as well as the ICML 55.3 Auditors' Standard Practice and Policies Manual.

The alignment of ICML 55.0 with other management systems and their associated standards was a priority in the development of this standard. Particular emphasis was placed on harmonizing this ICML standard with ISO 55000, ISO 55001, and ISO 55002 standards for asset management. As such, the structure and language of this standard have been generally harmonized with ISO 55001 (as appropriate) to assure strategic and operational alignment.

It should be noted that while this standard is intended to align closely with ISO 55001 and its subparts as amended, it is entirely the work product and exclusive intellectual property of the ICML organization and is neither explicitly nor implicitly endorsed by the International Organization for Standardization (ISO) or any other standards body.

1.3 Target Audience for this Standard

This ICML standard is intended for use by:

- Those who desire to improve the lubrication management practices of their lubricated mechanical assets pursuant to the realization of optimal organizational value as described by ISO 55001 and its amendments and subparts.
- Those involved in the establishment, implementation, maintenance, and improvement of a lubrication management system as a part of their physical asset management system as described by ISO 55001 and its amendments and subparts.
- Those involved in the planning, design, implementation, and review of lubrication management activities. These include local resources or outside service providers or advisors that provide contractual onsite support and/or services.

1.4 Benefits of this Standard

The adoption of ICML 55.1 requirements will enable the organization to achieve its objectives of effectively and efficiently managing its physical lubrication and lubricant asset policies, strategies, and plans.

The application of a lubrication management system for the organization's mechanical assets assures that these objectives can be achieved consistently and sustainably within the physical asset management plan over time.

1.5 Disclaimer

While every effort has been made to create a standard that is, in general, applicable to the lubricated physical assets located in a typical industrial facility, plant, or factory, it is not possible to anticipate or consider every application, machine environment, or circumstances associated with each of the potential applications of this ICML standard.

As such, this standard shall not be interpreted as an absolute substitute for principles of management that are based upon sound judgment in achieving reliable, safe, and economical asset performance. This includes those published guidelines and principles for maintenance and service provided by the mechanical equipment manufacturer.

Additionally, this standard does not account for applicable laws and regulations that could impact machinery operation or its maintenance.

> *"The adoption of ICML 55.1 requirements will enable the organization to achieve its objectives of effectively and efficiently managing its physical lubrication and lubricant asset policies, strategies, and plans."*

1.6 Lubrication Program Management – Overview and Terminology

1.6.1 Scope

This international standard is designed to provide an overview of a lubrication management system and processes applicable to the effective management of physical and nonphysical assets related to lubrication, its principles, and terminology.

This international standard can be applied to any type, size, and nature of industry or organization that employs mechanical equipment requiring lubrication.

This international standard is primarily intended for use in the management of both physical and nonphysical assets related to industrial lubrication that add value to the organization.

For the purposes of ICML 55.1, ICML 55.2, and this document ICML 55.0, the term " lubrication management system" is used to refer to a management system for lubricated assets. Although similar in purpose, this should not be confused with commercial software products that use the generic name Lubricant Management System to aid in managing lubrication practices, routine tasks, and inspections of mechanical equipment.

2

Lubrication Program Management

2.1 General

Factors that influence the types of assets an organization will require to achieve its lubrication program management objectives, and how they are managed, may include the following:

- Business type
- Purpose of the organization
- Operating context
- Plant layout and design
- Geography (plant location)
- Staffing compliment
- Regulatory requirements
- Financial constraints
- Organization needs and expectations
- Stakeholder needs and expectations

These factors must be considered when designing, implementing, maintaining, and improving a lubrication management program.

Effective lubrication management of mechanical assets delivers improved organizational control and governance. This translates into realized value through managed risk and opportunity to achieve a desired balance of cost, risk, and performance.

2.2 Benefits of a Lubrication Management Program

Effective lubrication management delivers a wide range of benefits to an organization, allowing it to realize and maximize an asset's value in many areas that include:

a. **Financial improvement**
   - Increased throughput due to reduced downtime
   - Increased asset reliability and life cycle
   - Reduced asset failure/repair costs due to reduced wear
   - Reduced lubricant inventory costs through lubricant consolidation
   - Reduced energy costs through engineered lubricant delivery
   - Reduced waste cost

b. **Risk management improvement**
   - Reduced financial losses (meeting targets/planned downtimes)
   - Increased health and safety
   - Positive client response due to improved meeting of production and maintenance targets
   - Minimized environmental and social impact
   - Reduced insurance costs and liability

c. **Improved service levels**
   - Increased output
   - Higher product quality
   - Planned maintenance downtime
   - Increased machine reliability/throughput

d. **Demonstration of compliance and social responsibility**
   - Reduced energy requirement
   - Reduced emissions
   - Recycled lubricants
   - Spill containment
   - Reduced waste
   - Meeting all legal, statutory, and regulatory requirements

e. **Improved efficiency and effectiveness**
   - Improved processes/procedures
   - Planning and scheduling of work
   - Achieving business objectives

2.3 Assets

An asset within an asset management system is an item, thing, or entity that has potential or adds actual value to an organization.

A lubrication asset is classified the same way and can be a lubrication system, a lubrication spare part, a lubricant, a lubrication work plan, a filter cart, a lubrication contract, a lubrication system drawing, a lubricant consolidation program, lubricator training, certification, etc.

2.4 Lubrication Program Management

2.4.1 General

The organization's management, employees, and stakeholders should work together to implement planning and controlled lubrication activities through policy, processes, planning, scheduling, and monitoring work activity to exploit opportunity and reduce risk.

Costs, opportunities, and risk should be balanced against lubricated asset performance to achieve organizational objectives.

2.4.2 Program fundamentals

Value: A lubrication program should be designed to add value to the organization and its stakeholders. Value can be tangible and intangible, financial, or non-financial. A value statement should include:

- How the program objectives align with the organizational objectives
- A life cycle-based approach to realize program value
- Established decision-making processes that reflect stakeholder needs and define value

> *"A lubrication program should be designed to add value to the organization and its stakeholders."*

Alignment: The lubrication program should align with and support the maintenance department program objectives and goals. The lubrication program should also align with the manufacturing/production objectives and goals as well as the corporate goals. Such alignment is often referred to as "line of sight."

Leadership: Leadership and commitment from all managerial levels are required. This is demonstrated through clear role definitions, responsibilities, and authorities. Before embarking on an ICML 55.1 certification journey, the workforce, management, and stakeholders should be consulted and subsequently remain involved throughout the process.

Assurance: To effectively govern an organization, assurances must be given so progress can be monitored and reported. Program and asset performance should be linked to organizational objectives. Process and performance monitoring indicators (KPIs) are used to monitor the program's continuous improvement and to isolate problem areas.

2.5 Relating the Lubrication Management Program to Asset Management

When an organization relies on mechanical equipment to produce its deliverable products, Figure 1 demonstrates how the lubrication management system is the cornerstone system that underpins and links to the asset management system, linking to all of the organization/corporate management systems.

> *"The lubrication program should align with and support the maintenance department program objectives and goals."*

The lubrication management system utilizes the asset portfolio that details all of the assets within the scope of both the lubrication management system and the asset management system.

The asset management system comprises interrelated or interacting elements that establish the asset management policy, objectives, and processes used to achieve those objectives.

Both the lubrication and asset management systems must align with each other and the corporate management systems. All elements are linked through the coordinated activity of the organization to realize value from the assets represented by the light blue asset management intermediary box.

Figure 2 shows the result of a major lubrication/tribology study[1] performed by Dr. Ernest Rabinowicz of MIT that found that 70% of all machine-bearing failures were due to surface degradation (wear); both a direct

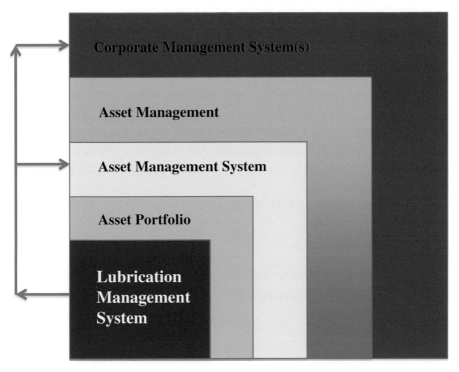

Figure 1 Corporate/system relationship.

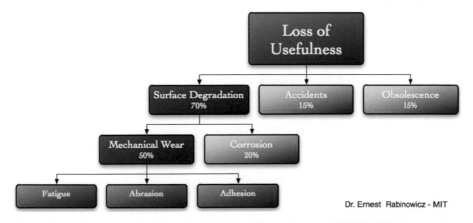

Dr. Ernest Rabinowicz - MIT

Figure 2 Lubrication-related bearing/machine failure. Courtesy ENGTECH Industries Inc.

and indirect result of ineffective lubrication practices that could have largely been avoided by the implementation of a lubrication management system.

2.6 The Lubrication Management System (LMS)

A lubrication management system is a collective series of linked elements designed to meet the audit requirements set out in ICML 55.1, as depicted in Figure 3. In addition, the LMS is designed to be a living system continually monitored and updated as new organizational plans, objectives, assets, policies, relationships, and performance evaluations trigger the need for adaptation and improvement.

The dotted box line in Figure 3 designates the boundary of the lubrication management system.

> *"A Lubrication Management System is a collective series of linked elements designed to meet the audit requirements set out in ICML 55.1."*

2.7 Organization Plans and Objectives

The organization usually comprises the clients who own the physical assets being lubricated and maintained, as well as other stakeholders within the organization or corporation whose objectives can be affected by a failure within the lubrication management system.

Objectives are results that need to be achieved within a timeframe or delivered at a defined level. For example, a production facility wants to achieve X manufactured pieces per week, per month, etc., at a defined level of quality, all affected by the lubrication process, or lack thereof. Organizational objectives set the context and direction with which the maintenance department must align or improve its lubrication management program to achieve the desired results.

2.8 Lubrication Management Plan (LMP)

Derived from organization plans and objectives, the strategic lubrication management plan is designed to specify how the organizational asset

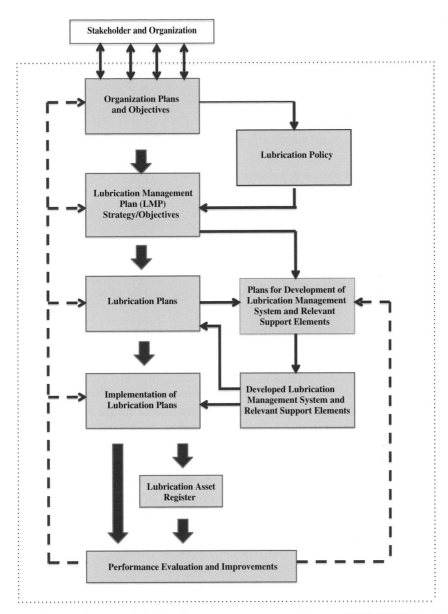

Figure 3 Key elements of a lubrication management system. Courtesy ENGTECH Industries Inc.

management objectives are to be converted into lubrication management objectives, and the approach used to develop the lubrication plans. It also serves to define the role of the lubrication management system (Figure 1) in supporting the achievement of the lubrication management objectives.

2.9 Lubrication Plans

Lubrication plans, sometimes referred to as lubrication management plans, document the activities, resources and timescale required for individual assets or groups of assets to achieve the desired strategic, tactical, or operational results.

2.10 Lubrication Policy

To aid the development of the lubrication plans, the lubrication policy (if in place) is also used. This document details the intentions and directions of the organization (corporation or department) as expressed by senior management.

2.11 Development Planning and Execution

Once the lubrication program direction is known, a lubrication operational effectiveness internal assessment can take place to understand the department's current state and assess what needs to change to allow the department to execute the lubrication plans and move to a future state ready for certification.

Often a lubrication expert is employed at this stage to assist the lubrication management department in understanding what change is required to prepare for an ICML 55.1 audit. Elements typically found to require attention may include a lubrication consolidation program, lubricant storage facility, contamination control program, inspection, oil analysis and application engineering upgrades/tune-ups, PM setup, maintainer/lubricator training and certifications, up-to-date lubrication asset registers, lubrication system mapping, performance evaluation system for KPIs, etc.

2.12 Implementation of Lubrication Plans

Implementation of lubrication plans requires an implementation project plan to be drawn up that will guide the organization from the current state to the future state.

Table 1 Typical ten-step ICML 55.1 certification process.

Step #	Action Item(s)	Yes	No
1	**Perform Third-Party Lubrication Readiness Review**		
	Lubricant Consolidation Program?		
	Contamination Control Program?		
	Dedicated Lubrication Staff?		
	Check List PMs?		
	100% Engineered Lubricant Delivery Systems?		
	Checklist PMs?		
	Oil/Grease Analysis Program?		
	Lube Routes?		
	Trained and ICML-Certified Lube Techs?		
	Dedicated Lube Transfer Equipment?		
	Lube System Mapping?		
	Current State/Future State Modeling?		
	Budgeting Plan in Place?		
	Lubrication Management Plan in Place?		
	Stakeholder Agreements in Place?		
2	**Review/Accept Non-Compliance Report**		
3	**Build Compliance Project Plan**		
	- Develop Implementation RACI Chart		
4	**Train and ICML-Certify Internal Lubrication Personel**		
5	**Train and Implement ICML Internal Audit Assessor**		
6	**Implement and Complete Compliance Project Plan**		
7	**Update Work Management and Reporting System**		
8	**Perform External Certification Readiness Audit**		
9	**Update as Required and Perform External ICML Audit**		
10	**Pass the Audit, Receive Certification and Fly the ICML 55 Flag**		

2.13 Performance Evaluation and Improvements

Once most of the LMS is up and running, an internal audit-ready assessment team can perform a mock audit to the ICML 55.1 Standard. These internal assessors can be certified through ICML-certified training organizations. Once any major or minor or non-conformity findings are rectified, the organization is then ready to be audited to the ICML 55.1 by an ICML-certified professional auditor and receive their official ICML 55 certification status.

2.14 Typical ICML 55.1 Audit Preparation Process

Once an organization has decided to implement a lubricated asset management system and move forward with certification, it must first undertake a "current state" evaluation of its lubrication program to understand its current state of compliance to the ICML 55.1 Standard requirements. This is best performed by a third-party lubrication specialist company who (1) understands the audit process, (2) has prior experience in the development of a lubrication management action plan, and (3) has a toolkit to facilitate the change process and reduce the preparation time to audit.

Table 1 details a ten-step certification process depicting a typical certification journey and some of the typical compliance areas to be addressed before certification takes place. If your current lubrication program is mature, many of the requirements may already be in place. Note: If you already have ISO 55001 certification in place, the process can be significantly expedited.

3

Terms and Definitions

For the purposes of this document, the following terms and definitions apply.

3.1 Asset

Item, thing, or entity that adds potential or actual value to an organization.

3.2 Asset Life Cycle

Stages involved in the life cycle of an asset.

3.3 Asset Portfolio

Assets that are within the scope of the asset or lubrication management system.

3.4 Asset System

Group of assets that interact or are interrelated.

3.5 Audit

A systematic, independent and document assessment process for obtaining audit evidence for objective evaluation against a set of requirement criteria defined in ICML 55.1.

3.6 Capability

The measure of capacity and ability for a person or organization to obtain its objectives.

3.7 Conformity

The application of knowledge and skills to achieve an intended result.

3.8 Continual Improvement

Recurring activity used to enhance performance.

3.9 Corrective Action

Action to eliminate the cause of nonconformity.

3.10 Documented Information

Information that is required to be controlled and maintained by the organization, including the medium on which it is stored.

3.11 External Audit Assessor

A person or persons employed by a professional licensed third-party auditing company who is/are trained and certified through ICML to perform ICML 55.1 certification audits (also see 3.13 below).

3.12 Incident

An unplanned event resulting in damage or loss.

3.13 Internal Audit Assessor

A person or persons employed by the organization who is/are trained and certified through ICML to perform internal assessments in preparation for the ICML 55.1 certification audit that will be performed by a licensed and certified third-party ICML 55.1 auditor (also see External Audit Assessor).

3.14 Lubrication Management

Coordinated activity to realize value from lubricated and lubrication assets.

3.15 Lubrication Management Plan (LMP)

Derived from organization plans and objectives, the strategic lubrication management plan is designed to specify how the organizational asset management objectives are to be converted into lubrication management objectives, as well as the approach used to develop the lubrication plans. It also serves to define the role of the lubrication management system (Figure 3) in supporting the achievement of the lubrication management objectives.

3.16 Lubrication Management System (LMS)

A collective series of linked elements designed to meet the audit requirements set out in ICML 55.1 as depicted in Figure 3. The LMS is designed to be a living system that is continually monitored and updated as new organizational plans, objectives, assets, policies, relationships, and performance evaluations trigger the need for adaptation and improvement change.

3.17 Monitoring

Action to determine the current status of a process system or activity.

3.18 Nonconformity

Nonfulfillment of a requirement.

3.19 Objective

A strategic, tactical, or operational result to be achieved.

3.20 Organization

Person or group of people operating with unique responsibilities, authority, and relationships to achieve specific objectives.

3.21 Organizational Objective

Overarching objective setting the context and direction for the organization's activities.

3.22 Organizational Plan

Documented information that specifies the programs to be used to achieve the organizational objectives.

3.23 Performance

A measurable result.

3.24 Policy

The intention and direction to be taken by the organization as expressed by upper management.

3.25 Process

Interrelated activity that converts activity inputs to outputs.

3.26 Requirement

Stated need or expectation, implied, or obligatory.

3.27 Risk

The effect of uncertainty on objectives. The effect can be a deviation +/−.

3.28 Stakeholder

Person or organization that can affect or be affected by a decision or activity.

Bibliography

1.   *Design, Friction and Wear of Interacting Bearing Surfaces* (1970) Rabinowicz, Dr. Ernest, Massachusetts Institute of Technology (MIT), MA, USA

Index

A
Audit preparation 13

E
External audit assessor 16

I
Internal audit assessor 16

L
Lubrication management plan 10,
    17

Lubrication management pro-
    gram 1, 5–6, 8, 10
Lubrication management
    system 1–4, 8, 10, 12, 15, 17
Lubrication plans 12, 17
Lubrication policy 12

O
Organizational plan 17

P
Performance evaluation 12–13